Awesome, Disgusting Careers

DISGUSTING FOOD JOBS

Stephanie Bearce

BLACK RABBIT BOOKS

Hi Jinx is published by Black Rabbit Books
P.O. Box 227, Mankato, Minnesota, 56002.
www.blackrabbitbooks.com

Marysa Storm, editor; Michael Sellner, designer and photo researcher

Library of Congress Cataloging-in-Publication Data
Names: Bearce, Stephanie, author.
Title: Disgusting food jobs / by Stephanie Bearce.
Description: Mankato, Minnesota : Black Rabbit Books, [2023] | Series: Hi jinx. Awesome, disgusting careers | Includes bibliographical references and index. | Audience: Ages: 8-12 | Audience: Grades: 4-6 | Summary: "Let readers explore the awesome, disgusting food jobs that keep their world running smoothly through witty, conversational text, fun facts, and critical thinking questions that'll have them laughing and learning"- Provided by publisher.
Identifiers: LCCN 2020030395 (print) | LCCN 2020030396 (ebook) | ISBN 9781623106812 (hardcover) | ISBN 9781644665480 (paperback) | ISBN 9781623106874 (ebook)
Subjects: LCSH: Food industry and trade-Vocational guidance-Juvenile literature.
Classification: LCC TP370.3 .B43 2022 (print) | LCC TP370.3 (ebook) | DDC 647.95023-dc23
LC record available at https://lccn.loc.gov/2020030395
LC ebook record available at https://lccn.loc.gov/2020030396

Image Credits

Alamy: Chico Sanchez, 16; Chris de Blank, 12; kevin nicholson, 4; miguel champi, 15; Miguel Cuenca, 15; Richard Griffin, 5; Dreamstime: Igor Zakowski, Cover; Kadirlookatme; 7; Sharon Kingston, 15; Shutterstock: Alica in Wonderland, Cover, 12, 21; Analia26, 2, 3, 16, 17; callumrc, 7; Charly Morlock, 19; ekler, 14; GoIF2532, 15; industryviews, 11; LoopAll, Cover, 1, 5, 6, 10, 20; Lopolo, 8–9; Memo Angeles, 10, 13, 20; michelaubryphoto, 12; Moriz, Cover, 23; Pasko Maksim, 6, 18, 23, 24; picoStudio, 9, 10; Pitju, 21; Plastic Man, 16; Rohit Dhanaji Shinde, 2, 3, 16, 17, 21; Ron Dale, 5, 6, 10, 20; Sararoom Design, 7; Serhii Bobyk, 11; totallypic, 5, 15, 19; Velveteye, 7; Vershinin89, 18–19; zcw, 12

CONTENTS

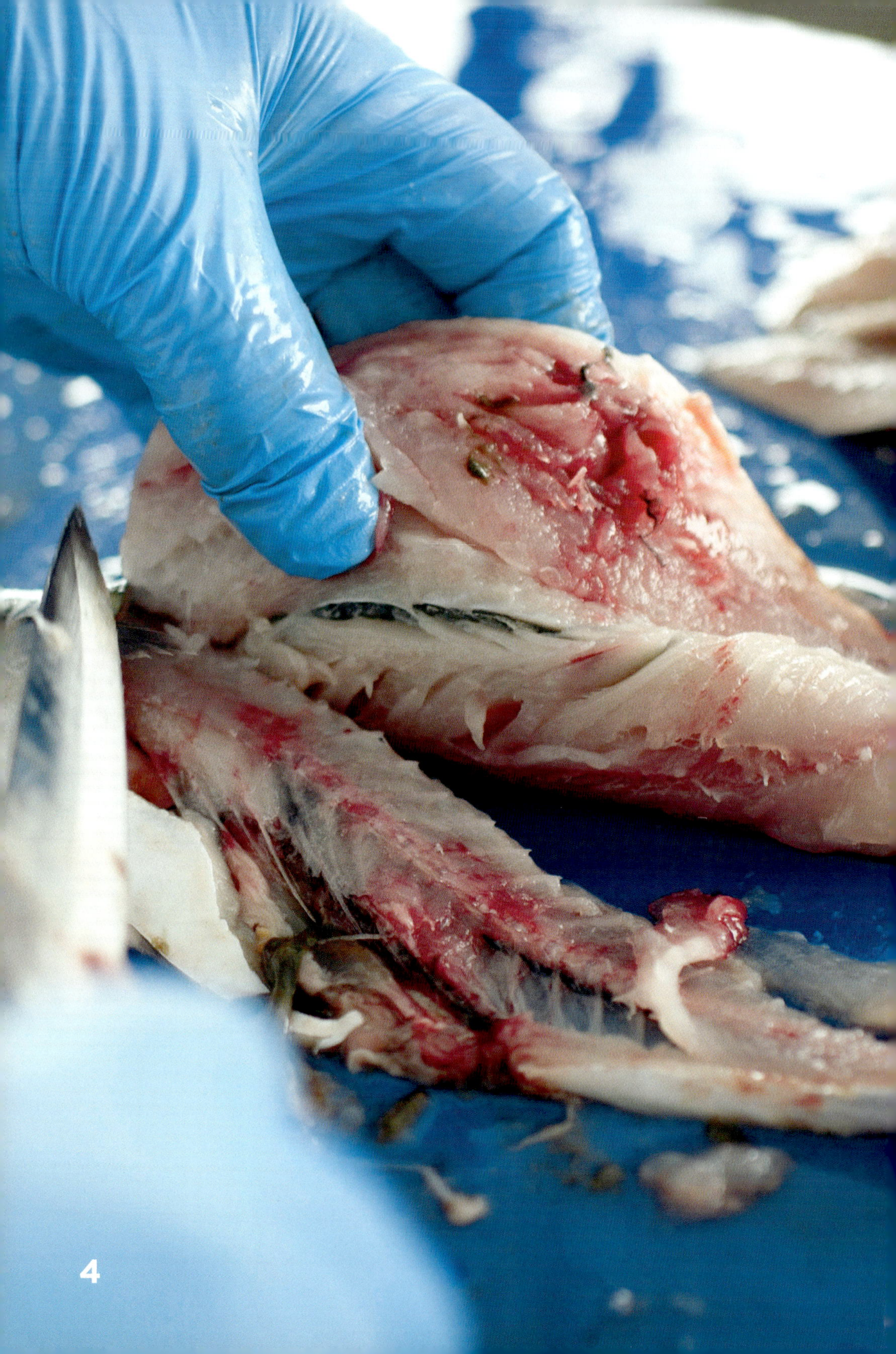
4

Chapter 1
FREAKY FOOD JOBS

Everybody loves to eat. Food is delicious. But working with food can be gross. It can be slimy. It can be smelly. Some food workers have to look for bugs. Others cut up fish and pigs. Some even work with poop. Do you have the stomach to learn more?

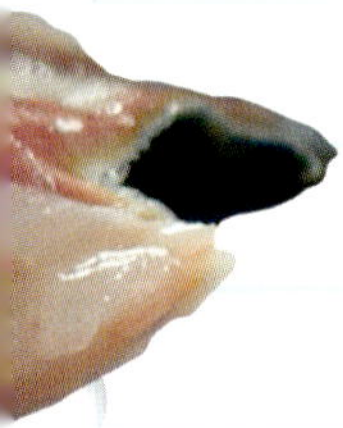

Chapter 2
THE JOBS

Restaurant Health Inspector

Bugs in your burger? Slime in your soup? Restaurant inspectors make sure these things don't happen. They go to restaurants and look for grease and gunk. They examine soda fountains and garbage disposals. Inspectors also search for pests. They make sure food is kept at safe temperatures. Inspectors force gross restaurants to close. The owners need to clean them before reopening.

Some inspectors say they can smell a cockroach problem. They describe the smell as nutty and oily.

greasy burner

food inspection

dirty kitchen

Professional Kitchen Cleaner

If a restaurant gets shut down, owners might call professional kitchen cleaners. These cleaners know how to fix a mess. They suit up in protective clothing. They use gloves and goggles. Then they pull grease out of pipes. They clean vents, ovens, and other surfaces. It's a nasty job. But it keeps people safe.

One of the dirtiest parts of a restaurant is a menu. Just think about how many people touch it.

Butcher

Hacking into bones. Chopping up hearts. That sounds like something from a horror movie. But it's actually the work of a butcher. A butcher's job is to cut up dead animals for meat. That meat is then packaged and sold. Butchers need to be strong. They also need to be careful. One wrong move could ruin a cut of meat. It could also mean losing a finger!

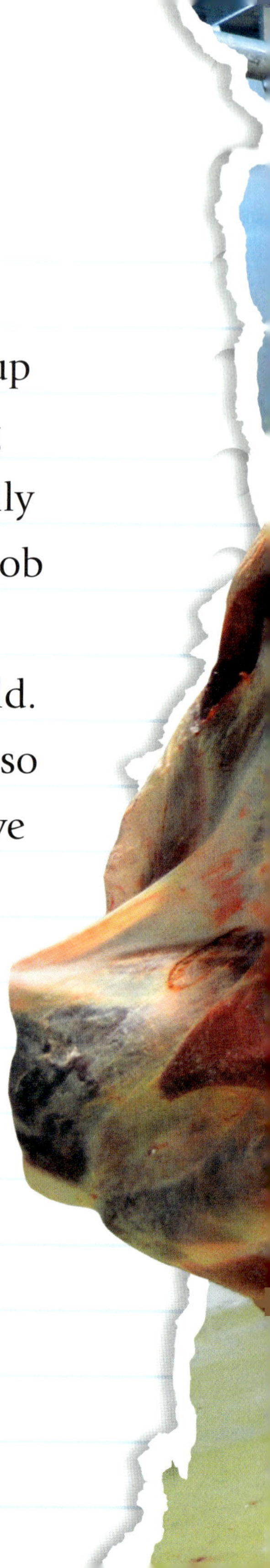

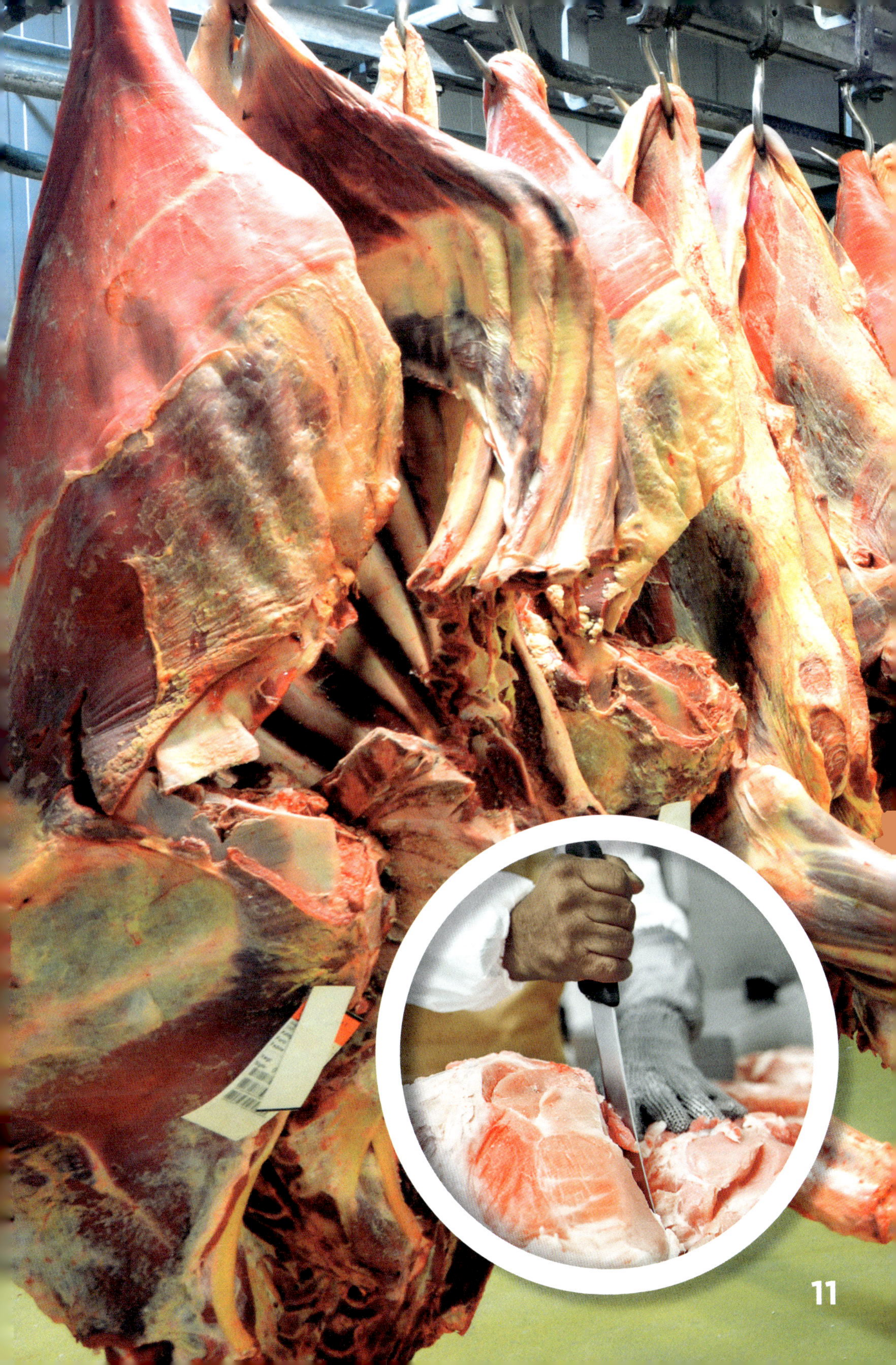

ground animal parts

stuffing meat

sausage

Sausage Maker

Sizzling sausages smell good. But making sausage is a disgusting job. Sausage makers grind up animal parts. They add spices. Then they mix the meaty mess together. Once the mix is ready, they take animal **intestines** and wash out all the blood and poop. They then stuff the meat into the guts. To finish up, they tie off the ends and cut the sausages apart.

Pig intestines are long. They can stretch more than 79 feet (24 meters).

Mushroom Farmer

Do you like mushrooms on your pizza? You might not after you read this. To grow mushrooms, farmers collect piles of **manure** and straw. They spray these piles with water and let them rot into **compost**. Once the compost is nice and steamy, they add **spores**. The farmers then stick the mess in a moist room. After weeks of waiting, mushrooms are ready.

Better wash those mushrooms before eating them. That dirt is probably poop.

WATER MIXTURE
COMPOST
MOIST ROOM
MUSHROOMS

Beekeeper

Human spit isn't very tasty. But bee spit is delicious. That's right. Bee spit is honey. And it's up to beekeepers to collect it. First, these workers suit up in protective gear. Next, they remove frames from the bee boxes, shaking off the bees. Beekeepers then use machines to spin honey from the frames. Beekeepers do more than collect honey, though. They must keep their hives free from disease. They need to get rid of any pests too.

Cheese Makers

Bacteria. You see that word and probably think of nasty stuff. But it's not all bad. In fact, people use bacteria to make cheese. Cheese makers start with a bunch of milk. They mix in bacteria and **rennet**. Over time, the milk thickens into a slimy, puddinglike mixture. The workers stir and rake this mixture. Then they drain off the liquid. The remaining gunk, called curds, is pressed into **molds**. After being molded, the cheese is left to dry or age.

Think making cheese with bacteria is gross? Some cheese is also made with **mold**.

molds
cheese
curds

Chapter 3
GET IN ON THE HI JINX

Sure, working with food can be nasty. But this hard work can also be fun and rewarding. And anyone can do it!

Many jobs give workers on-the-job training. They learn to do the jobs as they work. Other jobs, such as health inspector, require workers to take classes. These courses teach people about the tools they'll use. Students also learn the rules they must follow. Think a career working with food is for you?

Take It One Step More

1. Why are bacteria needed to make cheese? Research to find out.

2. Which job in this book do you think is the grossest? Why?

3. Some foods are made in disgusting ways. Do you still want to eat them after learning how they're made?

GLOSSARY

bacteria (bak-TEER-ee-uh)—a small living thing

compost (KOM-pohst)—a mixture of various decaying organic matter

intestine (en-TE-sten)—the part of the digestive system where most food is digested; it is a long tube made up of the small intestine and the large intestine.

manure (muh-NOOR)—solid waste from farm animals that is used to make soil better for growing plants

mold (MOHLD)—a hollow form in which something is shaped; it is also a growth of fungus on damp or decaying material.

oil (OYL)—a greasy liquid substance

rennet (REN-it)—a component needed to thicken cheese

spore (SPAWR)—a cell made by some plants that is like a seed and can produce a new plant

LEARN MORE

BOOKS

Mattern, Joanne. *Disgusting Jobs.* That's Disgusting! Minneapolis: Kaleidoscope, 2020.

Nickel, Scott. *World's Grossest Jobs.* Minneapolis: Lerner Publications, 2021.

Williams, Morgan. *Careers for People Who Love Cooking.* Cool Careers without College. New York: Rosen Publishing, 2020.

WEBSITES

Beekeeping Facts for Kids
kids.kiddle.co/beekeeping

How Cheese Is Made
www.highlightskids.com/explore/did-you-know/cheese

Weird but True!: Food
kids.nationalgeographic.com/explore/adventure_pass/weird-but-true/freaky-food/

INDEX